Science, Technology, Engineering, Arts, and Math: S T E A M.

STEAM is a way to take the benefits of STEM and complete the package by integrating these principles in and through the arts. STEAM takes STEM to the next level: it allows students to connect their learning in these critical areas together with arts practices, elements, design principles, and standards to provide the whole pallet of learning at their disposal. STEAM removes limitations and replaces them with wonder, critique, inquiry, and innovation.

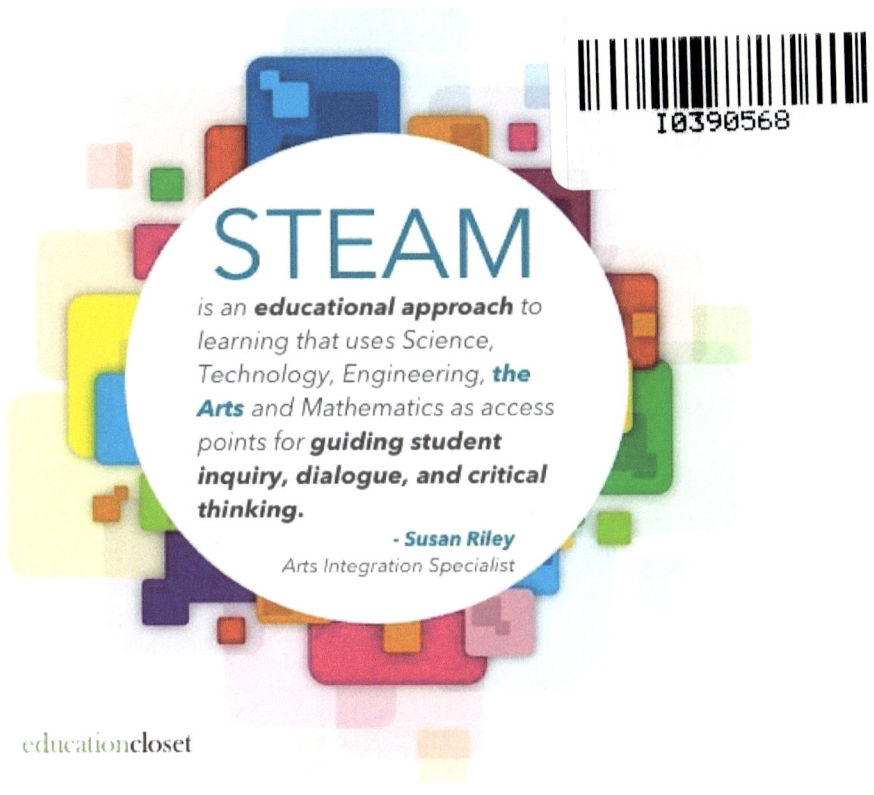

STEAM

is an **educational approach** to learning that uses Science, Technology, Engineering, **the Arts** and Mathematics as access points for **guiding student inquiry, dialogue, and critical thinking.**

- Susan Riley
Arts Integration Specialist

I0390568

educationcloset

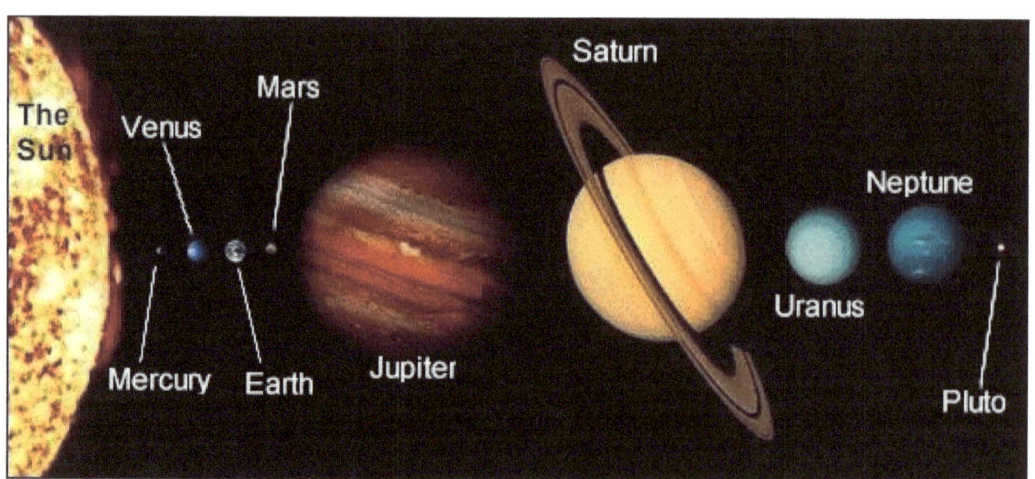

Mars Planet Profile

Mass:	0.642 x 10^24 kg (0.107 x Earth)
Equatorial Diameter:	6,805 km
Polar Diameter:	6,755 km
Equatorial Circumference:	21,297 km
Known Moons:	2
Notable Moons:	Phobos & Deimos
Orbit Distance:	227,943,824 km (1.38 AU)
Orbit Period:	686.98 Earth days (1.88 Earth years)
Surface Temperature:	-87 to -5 °C
First Record:	2nd millennium BC
Recorded By:	Egyptian astronomers

Table of Contents

The cover - shows the planet Mars, penned by Edgar Rice Burroughs in his famous novels as 'Barsoom'.

© 2019 J. Wayne McCain

ISBN: 9781081668877

The first *human* explorers to reach Mars will be further from Earth than any living person has ever been in recorded history. The very best day on Mars corresponds roughly to the worst day imaginable deep in the wilderness of Siberia! The Martian atmosphere is thin and not breathable. It is overbearingly cold, dusty, and windy. The thin atmosphere offers insufficient protection against radiation from the Sun and deep space. There is no unfrozen water on or very near the surface. There is no known flora or fauna to help sustain human life. A normal fire will not burn.

Mars is, in a word, desolate.

DEDICATION

This work is dedicated to Anniston High School - Administrators, Faculty, and Students during the 1968-69 school year, many of whom were participants in Project Aries 1. Without the trust and belief in these students by all concerned, a project of such magnitude could not have been attempted, much less completed successfully…

Anniston High School; Anniston, AL. has been a viable educational institution in Calhoun County Alabama since 1923. The original buildings and site were in use through the year 1970. In 1971, new buildings and facilities were opened on Woodstock Avenue a few miles from the original. AHS has been known over the years for it's forward looking approach to education and use of educational technologies. It is still considered one of the most advanced high schools in the state of Alabama. The original AHS campus shown above was closed one year after the completion of Project Aries 1.(lower photo credit Meg McCain)

PROJECT ARIES, THE NAME

The simulated Mars mission project was coined as Aries based on the Zodiac sign of the Ram for the anticipated mission timeframe of mid-March and April. Since at the time, additional simulated missions were being considered, the first was declared as Aries 1. [Author's note: Ares, the Roman name for Mars and the Roman God of War was picked by NASA for an early program that was to allow crewed-access to space beyond the Shuttle program.]

MARS, THE ASTRONOMICAL SYMBOL

Astronomical symbols for constellations, planets, and other astro events, date back in history beyond the medieval times. For Mars, the Roman God of War, the astronomical symbol is the shield and spear.

PREFACE

The 1968-69 school year was a turbulent one in the U.S. with Vietnam War protests, civil rights unrest, and a generation of youth becoming more disillusioned with 'The Man' and unfulfilled promises of the times. Amongst the confusion of tear gas bombs in city and suburban streets, there was one government activity that rose above the chaff: Project Apollo. Mankind had committed, via a young and vibrant U.S. President, John Fitzgerald Kennedy, to "send men to the moon and return them safely to the Earth" prior to end of the decade. JFK did not live to see the fulfillment of his dream, but on his shoulders and those of Dr. Wernher Von Braun, leading tens of thousands of other dedicated U.S. engineers and scientists, the goal was accomplished on July 20th, 1969 with only a few months to spare. Apollo was the culmination of a decade of space science and engineering research to meet the challenges of JFK's lofty goal – Projects Mercury and Gemini had paved the way and helped solve the unanswered questions that hadn't even been asked a decade earlier. These programs enriched the imagination of thousands of American youth to 'reach for the stars' in emulation of these great accomplishments. This overview tells one such story, where over 30 high school students attending Anniston High School (Alabama), living only a hundred miles from Marshall Space Flight Center where the Apollo Saturn V and Von Braun's Rocket Team was based, turned away from less productive distractions of the times to dream a dream of Mars…and how we as a society might get there. We're still dreaming; still waiting.

THE REFERENCE MISSION

The primary Project Aries participants were Richard Austin, James Garrett, Sam Martin, Wayne McCain, Wallace Price, and Jim Young; all were Anniston High Seniors with the exception of Garrett, Price, and Young. These students had been developing and presenting individual annual science projects since junior high with regular success. In September 1968, the idea of a collaborated, joint project was introduced within the group. Martin, McCain, Young, and Garrett had previously entered projects with space-related topics ranging from model rocketry research to amateur astronomical photography. With the continued success of NASA's Apollo program and the predicted moon landing event in sight, the idea of a joint space project caught hold. Since all of the group had been interested in and were following the progress of Apollo, it was a short stretch to choose a space mission simulation. Just duplicating what NASA was currently doing wasn't deemed as 'innovative' by the group. Instead, it was decided to look beyond the moon landing program and choose Mars as the topic. The concept represented what appeared as the next logical step beyond the moon. So, Mars was the destination, what were to be the particulars? A full-scale Reference Mission needed to be defined to allow scaling a doable simulation for the joint science project.

Figure 1 – Apollo 8 Crew Lovell, Anders, and Borman (NASA Photo).

INSPIRATION for the Aries 1 Reference Mission was taken from one of Apollo's most dramatic accomplishments prior to the Eagle's landing on the moon in July. That was Apollo 8 in December 1968 wherein Jim Lovell, Bill Anders and Frank Borman (Figure 1), only the second manned Apollo mission and the first to fly aboard the Saturn V, circled the moon on Christmas Eve. The crew of Apollo 8 took turns reading passages from the Bible, most notably Genesis, that captured the attention and awe of many Earthlings. It was also during the Apollo flight that the famous Earth Rise photo was taken (Figure 2). Although the mission was not originally designed to orbit the moon, the opportunity presented itself and allowed several unique opportunities to validate Apollo hardware and mission planning. The risks were considered acceptable and the Apollo 8 mission to send the first humans to orbit the moon was a resounding success. Everyone waited with baited breath for what was to follow.

Figure 2 – 'Earth Rise' famous photo from Apollo 8 - an isolated blue spec in a dark ocean of space [NASA].

The Aries planning group figured that it was perfectly logical to conduct a similar 'around and back' Mars flight to prove out spacecraft systems, the trajectory acceptability, etc. In addition, the Aries 1 mission was to include photographic recon of potential landing sites conducted from the main spacecraft and deployment of a small lander to gather additional data at the most promising landing area. A three-man crew was again selected based on the Apollo program. No extravehicular activities (EVAs) were planned. The Aries 1 Reference Mission profile is shown in Figure 3 and was to be accomplished based on known and emerging technologies of the day. Unbeknownst to the Aries 1 group, Dr. Von Braun had also been thinking of what was to follow after the Apollo program was completed and had NASA contracted with Boeing and others to update his earlier Mars calculations (originally made in the 1950's) to take into account technology advancements already in hand and those that he considered on the near horizon. That study, which was not available publically at the time, would have been extremely valuable to the Aries planners. Never-the-less, the group forged ahead to lay out a reasonable Mars orbital mission scenario.

As shown in Figure 3, the main attributes of the Reference Mission include final assembly of the spacecraft in Earth orbit. It was envisioned that sections of the craft would have been launched into orbit using the Saturn V appropriately truncated (e.g. minus final stage and Apollo crew components) as a heavy launch vehicle (HLV). The spacecraft baseline would utilize nuclear heated LH_2, ionized and accelerated to exhaust velocities based on electrical power generated by the reactors (envisioned at the time to be SNAP 8 or a similar configuration). In the order of three to six separate launches were envisioned for transport of the modules into orbit where they would be assembled by the crew using space tugs and automated assist units. The LH_2 tank(s), reactor cores, and engine units would likely each be a separate Saturn V launch. The occupied crew module would be the final element to be launched. The overall vehicle once assembled and checkout completed, would await the proper timing for launch into the proper transfer orbit to rendezvous with Mars. There are several possible scenarios for launching a manned vehicle to Mars, it's not a predetermined decision. The parameters involved include transit time, stay within the vicinity of Mars, required fuel, vehicle complexity, and mission reliability. The longer missions beg for higher reliability systems, thus a classic tradeoff situation.

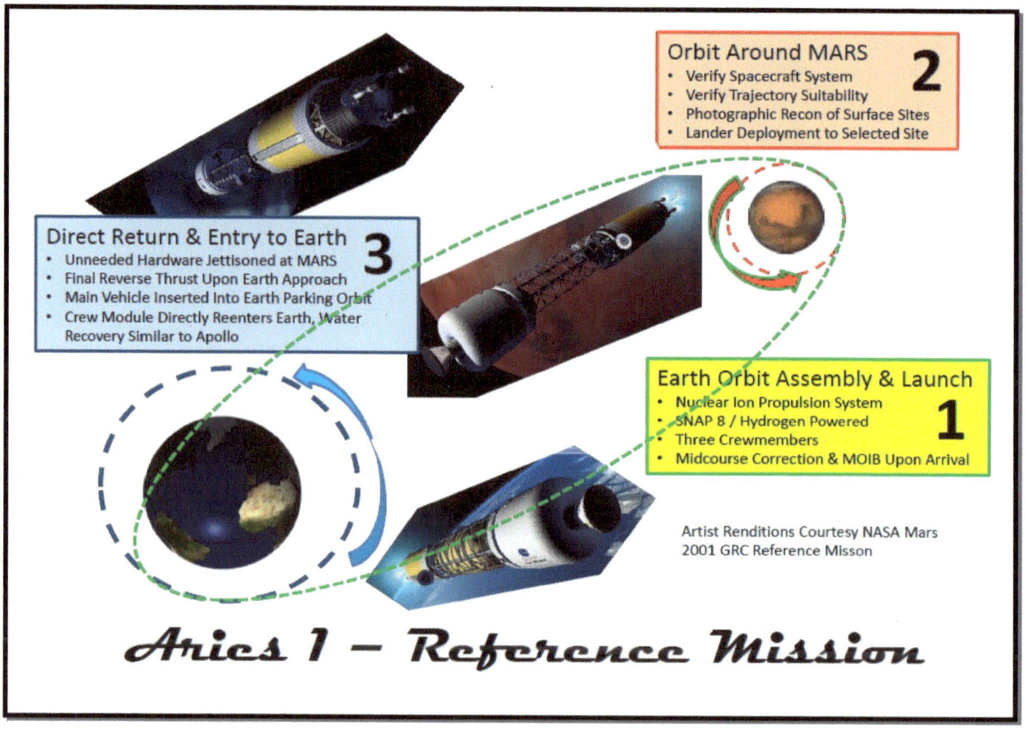

Orbit Around MARS **2**
- Verify Spacecraft System
- Verify Trajectory Suitability
- Photographic Recon of Surface Sites
- Lander Deployment to Selected Site

Direct Return & Entry to Earth **3**
- Unneeded Hardware Jettisoned at MARS
- Final Reverse Thrust Upon Earth Approach
- Main Vehicle Inserted Into Earth Parking Orbit
- Crew Module Directly Reenters Earth, Water Recovery Similar to Apollo

Earth Orbit Assembly & Launch **1**
- Nuclear Ion Propulsion System
- SNAP 8 / Hydrogen Powered
- Three Crewmembers
- Midcourse Correction & MOIB Upon Arrival

Artist Renditions Courtesy NASA Mars
2001 GRC Reference Misson

Aries 1 – Reference Mission

FIGURE 3 – The Aries 1 Reference Mission Main Attributes.

Figure 4 shows two of the popular approaches for flight from Earth to Mars. The left plot shows the straight-forward approach Conjunction mission using two Hohmann transfers between the two planets. An absolute minimum energy orbit would require about a 258-day transit each leg, but this could easily be shortened to around 180 days without requiring too much additional fuel. Studies done by Dr. Robert Zubrin and others show that this type of trajectory would require approximately 5 km/s departure velocity from Earth and would yield about a 180-day transit each way for a total mission time of 2-years for a Mars 'free return' mission. The trajectory depicted on the right of Figure 4 is referred as the 'Venus Fry-by' by Dr. Zubrin and is an approach to accomplishing an Opposition type trajectory. An Opposition trajectory puts Earth and Mars on the same side of the Sun at departure whereas the Conjunction genre mission puts Earth on one side and Mars on the other. In general the Opposition class orbit is considered less attractive due to longer transit times, shorter stay times near/on Mars, and mission reliability. Figure 5 shows standard Earth-Mars transit parameters with ΔV requirements for each type. Most NASA baseline planning considers some form of Earth-Mars trajectory scheme as was envisioned for the Aries 1 Reference Mission.

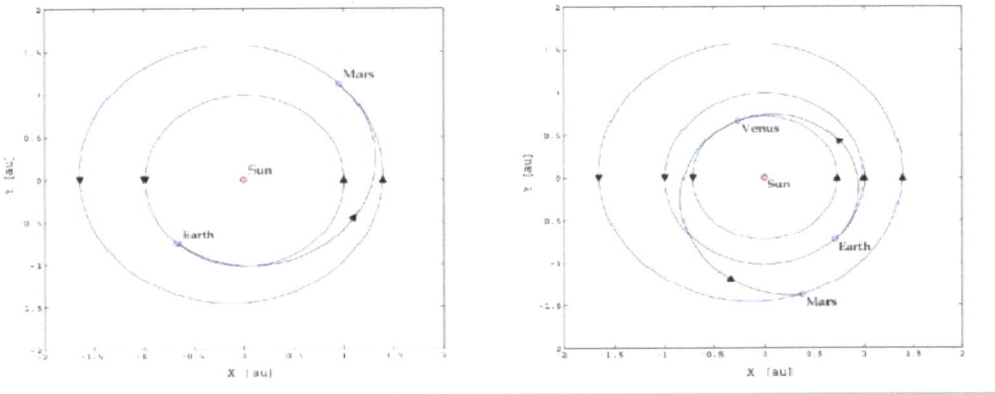

FIGURE 4 – Popular Orbital Schemes for Traveling Between Earth and Mars. The Left-Hand View Is the Chosen Approach Which Represents A Conjunction Hohmann Transfer Orbit.

Once the Reference Mission was set, it was necessary to translate that real mission into something that could be simulated within the time remaining in the school year. Negotiations with the AHS Faculty and Administration yielded that approximately two weeks was the maximum allowable time for the actual mission part of the project. After meetings between the Administrators and the involved students/parents, the timeline was set: a 16-day mission from launch out of Earth orbit until recovery of the crew at mission's end. It was noted that nothing like this had ever been done in the history of the Anniston school system so we would have lots of attention focused on the project and all activities to help insure the safety of all involved.

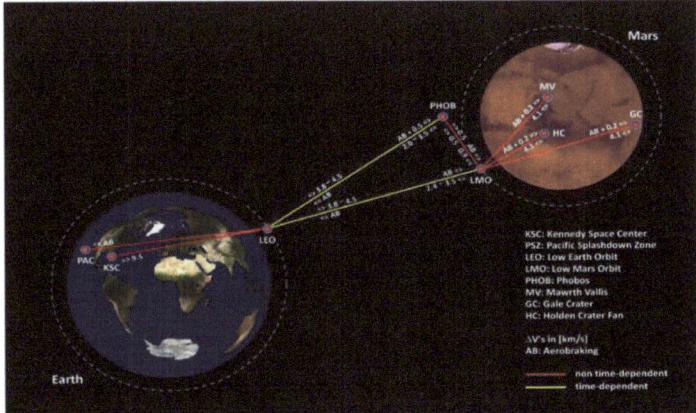

FIGURE 5 – Earth-Mars Standard Transit Parameters [MIT].

THE SIMULATION, CREW, AND SUPPORT TEAM

With the Aries 1 mission defined as a 16-day Mars Orbital Recon, the Team set about assembling the personnel and hardware to accomplish the simulated flight. Table 1 shows only a portion of the students and Faculty who participated in the project, but covers the major functions of the mission.

The three Flight Crew members were chosen primarily based on interest in going and their receiving permission from family and Faculty to do so. Wayne McCain, as senior member of the flight crew candidates, agreed to act as overall Mission Commander. Sam Martin who had recently won a science fair competition with his Micro-Biological Laboratory (MBL) spacecraft model (with functioning sensors and recorders) would be Science Officer (yes, the original Star Trek and Spock were fresh in everyone's mind!), while Wallace Price agreed to act as Pilot, essentially Mission Sub-Commander.

The Ground Crew was headed up by Richard Austin as Flight Director. Richard exhibited the oversight skills and was recognized as an authority figure amongst the group so was a natural for the position. Dennis Haynes was the Aero-Mechanical Lead, primarily because his Dad had offered to share some spare avionics parts, switches and circuit breakers, etc. from a salvaged Beechcraft Bonanza! Ken Mok, whose family owned a local Chinese restaurant, was known for his work with digital computer circuits (while the school system had a main-frame that they allowed us to use for our trajectory calculations, personal computers were many years in the future) served as our computer gurus. Jim Young, a long-time friend and short-wave listening enthusiast was Mission Control Director/communications specialist while James Garrett was in charge of astrophotography and photography in general. Heading up the Planetary Photography Lead role, was Sally Wallace, Assistant Flight Director, who was also in charge of the mock-up Recon Camera control panel. Sally, a very cute, spunky, sharp student in math and science, was on the cheerleading squad as well. Sally, along with all the other girls involved, was instrumental to the Project Aries success.

The AHS Administration and Faculty were outstanding in their acceptance and cooperation of the endeavor. After the initial meeting in December, the Aries Team was required to submit a formal, written proposal. We responded within two weeks with a 50-page, type-written plan (I still have my Royal 440). After studying our plan, the go-ahead was given. We were amazed to have access at all hours of the day, 24-7 for months leading up to and during the actual simulated flight. Mr. Clifford Smith, our infamous hard-nosed Principal and his Vice Principal Mr. Norman Messina, were actually very cooperative. In addition, the Faculty went WAY beyond their normal call-of-duty in advising and helping during the project. For instance, Terry Niblett, an avid scuba diver, accompanied the flight crew and several support personnel to a scuba training session at the deep and cold waters of Morrison Springs, FL where we experienced the feeling of simulated 'weightlessness'. There were many similar examples such as Mrs. Slate's offering up her foreign language laboratory as Mission Control. All hardwired com channels from within the Aries 1 spacecraft were run outside the second floor room where the craft was built and into the basement where the L-Lab was located. She frequently entertained the student crews with refreshments at her home which was only about two blocks or so from the school. Several other Faculty also took the Aries Project students under wing.

TABLE 1 – ARIES 1 CREWS & ADMIN/FACULTY, PARTIAL LIST

Flight Crew	Role	Age
Samuel Martin	Science Officer	18
Wayne McCain	Mission Commander	18
Wallace Price	Pilot	17
Ground Crew	**Role**	**Age**
Richard Austin	Flight Director	17
Sally Wallace	Assistant Flight Director	17
Ken Mok	Computer Specialist	18
Jim Young	Mission Control Director	16
James Garrett	Astro Photography	16
Dennis Haynes	Aero-Mechanical Lead	16
Chris Worsham	Security Chief	17
Sam Wheat	Analyst	17
David Gray	Chemical Engineering	16
Admin/Faculty	**Role**	
Clifford Smith	Principal	
Norman Messina	Assistant Principal	
Mrs. Mary Davis	Physics	
Terry Niblett	General Science	
Mrs. Clay Slate	Language/Communications	
Mrs. Judy Andrews	Human Anatomy/Medical	

Figure 6 shows part of the ARIES Team undergoing Scuba instruction with Faculty member Terry Niblett to enhance the Flight Crew training experience. Terry was unfailing in his encouragement and assistance to the students of Project Aries. Morrison Springs was a popular cave diving spot and offered ultra-clear, albeit cold, fresh spring water, which was slated for distribution as bottled water. It was announced during the Aries Team visit that the springs would be closed to divers. It has since reopened and is as popular as it ever was. Underwater training is still used by NASA to accustom crew to the phenomenon of Zero-G.

FIGURE 6 – ARIES 1 Scuba Training At Morrison Springs, FL Spring of 1969. L-R: Wayne McCain, James Garrett, Terry Niblett (back to camera), and Sam Martin.

Permission was granted to use one of the 2nd floor chemistry lab storage rooms as the location for building the Aries 1 simulator. Figure 7 shows the basic two-deck layout. Fortunately for the project, the AHS rooms had 12-foot ceilings which made construction of the configuration shown possible. The bottom deck was basically the living/personal hygiene quarters set up to accommodate three crew members. The upper deck was the spacecraft flight control cockpit.

On Deck 1, bunks were double-stacked on one side with a work area above the single bunk. A small galley was positioned at the end of the port side (looking aft toward the ingress-egress hatch). The Team included several members from the Home Economics Club, who together with the Future Physicians Club were in charge of identifying and packaging primarily freeze-dried meals, supplemented by canned vegetables and some canned meat. Military MRE's were used to finish out the complement of space food. A small water dispenser with a metal spout that would protrude into the packaged freeze-dried meals (chicken and tuna salad) was fashioned for use. All things considered, the space food diet was very acceptable and we were guaranteed to have a balanced and nutrient rich diet by the Team. On the other end of the spectrum was the chemical toilet. No commercial units were available for any price that we could afford, so we designed and built our own that included a feces breakup blade and manually administered chemicals. The toilet turned into one of our biggest worries beforehand although it performed satisfactorily during the mission (thanks David Gray). Storage containers similar to Tupperware® were used to organize food and other stores and were crammed in almost every otherwise unused space on the craft, both decks.

A custom built wooden ladder connected the decks. Deck 2 was the spacecraft cockpit where simulated control panels were constructed for all major systems (e.g. nuclear power, propulsion, navigation, communications, recon camera, and deployable surface lander). All of these designs were think-tanked and were as reasonably representative of their function as we could make them. For instance, for the navigation panel we fashioned a 4-digit electromechanical computer of sorts that was used for basic math and to fire the nuclear-ion engines during flight maneuvers. We used analog volt and amp meters for the power systems since digital readouts were not yet available at the consumer level although we did have several surplus nixie-tube readout devices that were used in Mission Control and outside the simulator.

Figure 8 shows more detail as to how the two decks were laid out. Construction was basically with 2x6 and 2x4 studs and plywood recovered from a local lumber yard discards pile. We made several pickup truck loads from the facility during the course of the Project. We were also given access to the State Surplus Center that was located in Gadsden, AL about a 30 minute drive from Anniston.

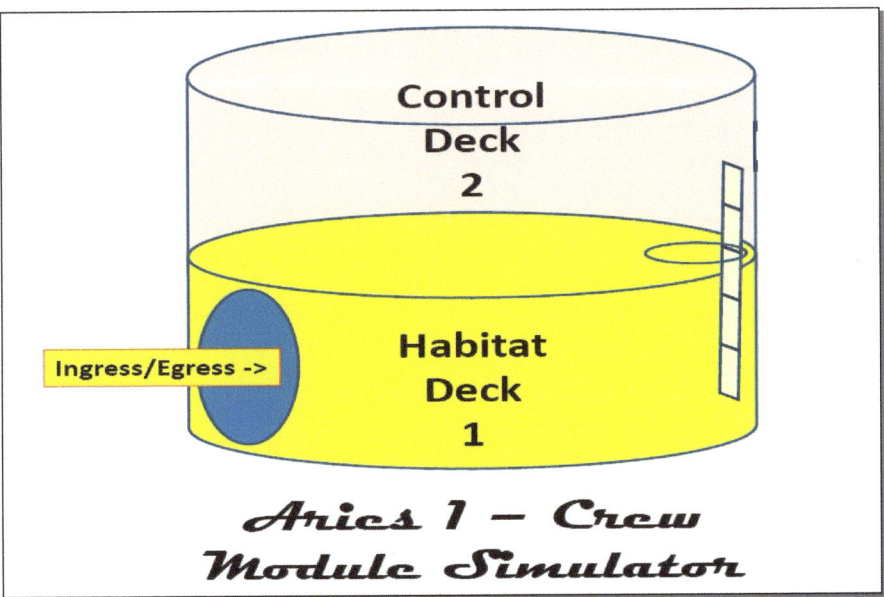

FIGURE 7 – Basic Two-Deck Layout of The ARIES 1 Spacecraft Simulator.

There the Team was able to load up a pickup truck with all kinds of electronic parts and laboratory equipment ranging from pen recorders to oscilloscopes at literally 10 cents per pound. For the Aries Team geeks, it was like an unattended candy store – full of chocolate!

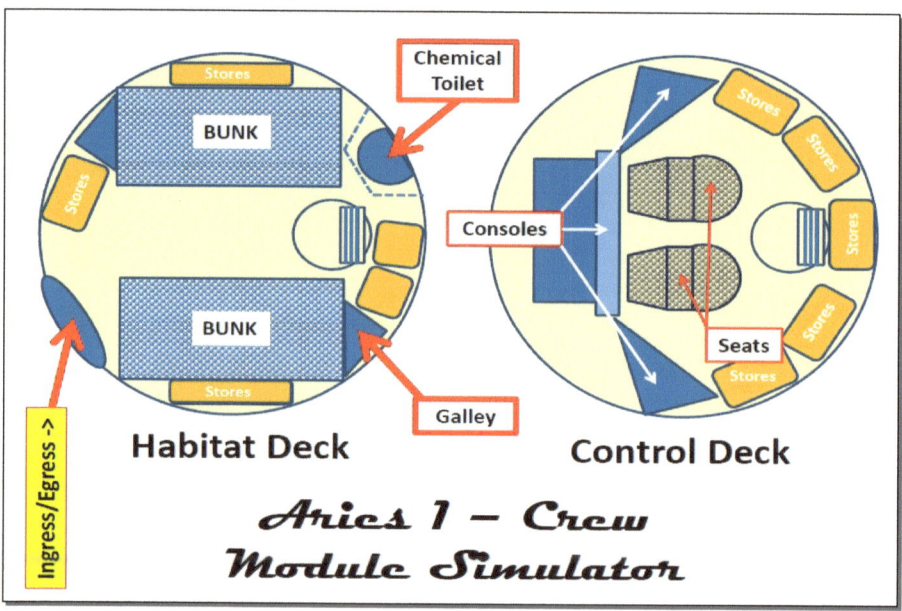

FIGURE 8 – Aries 1 Deck Detailed Layouts.

Approximately three months were required for construction and checkout of the simulator. The Project Aries Team had various groups working simultaneously (fast-tracking!) on tasks almost around the clock. Unprecedented access to school facilities was granted and some of the Team frequently worked through the night on weekends and on some school days. Once the Hab Deck was completed, it was not unusual to find Team sleeping through a class during normal school hours because they had just pulled an all-nighter (did I say that out loud?!). We had originally planned to launch during March, but as the construction drug out a bit, the date slipped into April. All aspects of a normal space mission were addressed to the extent possible. Crew helmets fashioned from motorcycle headgear and bubble visors, red jump suits, and ID Badges. The details became mind boggling near the T-0 date. Up to the simulation launch date, we had stayed under the radar of local newspaper, radio, and Birmingham TV. It was just as well since the Team didn't have the time or resources to deal with it and AHS, at that time anyway, had no formal public relations person of any kind to coordinate such activities. We were ready on the predetermined date for Launch. At a few minutes after 8:00 AM on April 8, 1969 Aries 1 officially left Earth orbit for Mars! The hardest parts of the Project were behind us, but more adventures lay ahead!

THE MISSION

The Anniston High School public address system carried live coverage of the Aries 1 launch thanks to the expert wiremanship of Jim Young and his understudies. All three crew were settled in their respective launch positions as the now-warmed up and ready SNAP 8 reactors were humming and high voltage was ready to accelerate the plasmatized Hydrogen atoms and sling Aries 1 out of Earth orbit and on the first leg of a Hohmann transfer to Mars. The Nav computer, basically acting as a countdown clock in this application, click-clacked, click-clacked down towards zero time to the Transfer Orbit Insertion Burn (TOIB). The big blue-ruled computer paper, donated by the downtown City school system computer center, provided the visual cues. Four separate reactors and engine controls were enabled, armed, and awaiting the final fire command from the computer. Jim Young, the voice of Mission Control, was relaying status and other systems data to and from Aries from the basement L-Lab. The borrowed video cameras and recorders were running. The simulation actually became rather tense as the seconds clicked down.

Counting: three - two - one – zero. Right on cue, four red engines running indicator lights came on at exactly **T - 0000** seconds. The craft shuddered. A gradual buildup of imagined acceleration was pushing against the crew and objects within the Aries 1 crew compartment. A determined whine was building as outside the spacecraft, the star scape slowly was changing as Aries headed towards Barsoom, some 1.5 AU away as the space crow flies. Farther and farther below, the brilliant blue orb that was the cradle of all mankind slowly slipped into the void. As the gentle acceleration built on the velocity vector in a cumulative, predetermined way, Aries headed somewhere that humans hadn't visited, at least in known history.

It helps to have a good imagination for simulations to seem more real. Of course the months of study and training didn't hurt either. A major objective of the Aries 1 mission was to show that astronauts could endure long term space voyages and perform adequately complex tasks. The Reference Mission for Aries 1 would have taken years to complete in real time, but for now, the Aries 1 crew traveled faster than man had ever traveled to be back home in time to finish those English term papers, the final physics and pre-calculus exams, and hopefully, walk across the graduation stage in late May, looming only a month away.

It seemed ironic that this mission had been inspired by Apollo 8, when at about 2 days in, Sam Martin our Science Officer began vomiting and experiencing other stomach flu symptoms just as Frank Borman did initially on the Apollo mission. Unfortunately, Sam also seemed to have a mild case of claustrophobia, that when combined with the other flu symptoms, resulted in a noticeable state of delirium. So, as much as the whole Team hated to, Sam was ejected in a lifeboat module and picked up by a rescue vehicle. Sam recovered nicely back on Earth and in a few days he was providing invaluable support from Mission Control. On Apollo 8, NASA only learned of Borman's dilemma a day or so after the fact because of communication restrictions, but he recovered enroute and neither of the remaining Apollo crew came down with his illness. After thoroughly sterilizing the innards of the craft, smooth sailing lay ahead for Aries. Routine course corrections, comparing actual with planned trajectory, and astronomical observing occupied crew time. The next major milestone, a deceleration and capture into Mars orbit, came off without a hitch.

The highlight of the Aries 1 mission plan was hours of planetary reconnaissance and deployment of the surface lander after the photography had been analyzed by scientists on Earth. The lander was to use aero braking to enter the Martian atmosphere and active guidance to the selected landing zone. Sally Wallace was on-hand steadily at Mission Control to scrutinize the photographic results as well as performance of the crew with her baby, the recon camera. At the designated time, the lander was released and successfully entered a trajectory to the Martian surface where higher resolution photos were taken and soil samples analyzed using Sam Martin's MBL technologies. It seemed like hardly any time had elapsed when the engines were once again lit to propel us on an elliptical arc towards home.

At shortly after 8 AM on April 24th, the Aries 1 was captured into an Earth orbit. The Crew Module reentered and was recovered by an anxiously awaiting Ground Crew as shown in Figure 9. Table 2 summarizes some of the tangible aspects of the Reference Mission as they were addressed in the simulated flight. A major accomplishment was that all of the Aries Team were still friends and on speaking terms!

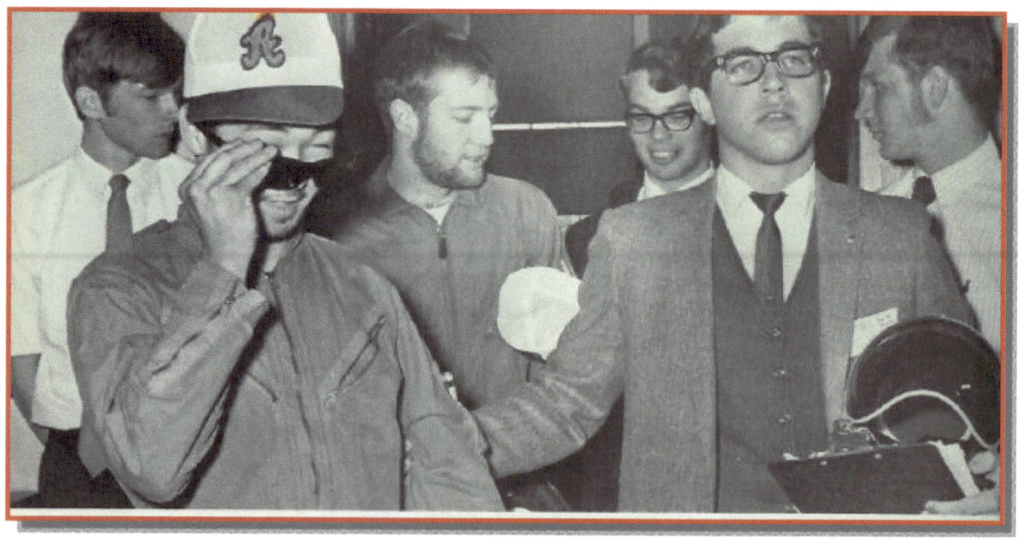

FIGURE 9 – Aries 1 Crew Lands At AHS: L-R, Steve Stewart, (Wallace Price, Wayne McCain, Flight Crew), Chris Worsham (Aries Security Chief), Sam Martin (Flight Crew), and Ralph Murray.

Immediately, the two Aries 1 crew, McCain and Price, were rushed to a waiting medical Team (Future Physicians again), then once cleared (both passed with flying colors), they were escorted to a school-wide assembly where 2001 Space Odyssey was expertly performed by the AHS Concert Band. Several Project Team members, including Martin, McCain, and Worsham were AHS Band members. It was a spectacular culmination to over 6 months of work; lots of blood, sweat, and tears for all concerned. Following the assembly, the crew was allowed to shower and change into civilian clothes in time for an official reception complete with Aries 1 cake and punch. All of the Ground Crew attended as did many Faculty and other students. Several days of debriefing ensued before most of the Project members had to face the harsh reality of needing to complete about 3 months of school work in only a month. No rest for the weary…

It must have been a bit similar to the feelings of Neil Armstrong, Buzz Aldrin, and Michael Collins after returning to earth from their successful first landing on the moon. **Now what?** So much energy, time, and effort had been devoted to Aries for a good portion of the school year that was difficult to get back into gear of a routine that demanded immediate attention. The usual determination prevailed, however, and most all assignments were completed, turned in, resulting in those graduating seniors walking as scheduled. Accolades were received from then Governor Al Brewer, local dignitaries, newspaper salutes, and the most coveted of all, a personal letter from Dr. Wernher Von Braun at Marshall Space Flight Center, Huntsville.

Most of the last week of school was spent dismantling Aeries 1. It came apart a lot quicker that it went together. The chemistry storage and L-Lab were returned to normal for the 1969-1970 school year which was to be the last one at the original Anniston High School campus. A new campus and a new world awaited - a world where mankind was no longer restricted to our place of origin, the planet Earth.

TABLE 2 – Reference Mission Issues Addressed by Aries 1 Simulated Mission

MISSION CONCERN/ISSUE	REFERENCE MISSION	ARIES 1 SIMULATED MISSION
Mission Trajectory Design	Hohmann Transfer Conjunction Orbits	Orbits Calculated, Times Truncated To Fit Simulation Timeline
Propulsion System Selection	Nuclear-Ion Propulsion	Simulated Propulsion Control Panel
Power Technology Selection	SNAP-8 Fission Reactors	Simulated Power Control Panel
Design of Diet & Storage of Food	Dietician Plan Low Carb	Custom Planned and Prepared
Adequate Environment Control	Self-Contained ECU	Modified HVAC Direct System
Water Storage/Generation	Recycle Potable Water	Fresh Water Inlets Provided
Human Waste Disposal/Recycle	Space 'zero-g' Toilets	Custom Chemical Toilet System
Communications & Delays	Tracking & Com Network	Simulated 20 min delays at Mars
Crew Mission Task Details	Detailed Mission Planning	Some Detailed Planning
Trajectory Corrections, TOIB, MCCB, MOIB, EOIB	Computer-Controlled Course Corrections	Simulated Corrections & Orbit Insertion Burns, Clock Controlled
Crew Module Design Rationale	2-Deck Hab & Controls	Built Functional 2- Deck Simulator
Radiation Hazards Avoidance	Provide Shielded Compartment For Crew	Layout of Crew Module With Stores Shielding Around Crew

EPILOG

This brief overview has only scratched the surface of all the planning, hard work, and outcomes of the Aries 1, Simulated Mission to Mars. For the group of students, the entire experience was certainly monumental if not life changing. We all experienced a level of organization, teamwork, and cooperation that could not have been achieved outside of an organized and sternly coached sports team. While Project Aries required grueling work and dedication, it was a labor of love and lots of fun too.

Subsequently on Sunday, July 20th, 1969 Neil Armstrong and Buzz Aldrin landed the Eagle Lunar Excursion Module (LEM) on the Sea of Tranquility. Shortly thereafter, Neil stepped foot on another world. Life on Earth has never been quiet the same since.

While some of us went on to pursue our dreams of traveling to Mars in various ways, I believe that we opened up a new partition of thought for ALL members of the Project that would not have occurred without the experience of Aries 1. Much later, it was revealed that Dr. Von Braun was updating his Mars exploration plans at the very time Aries 1 was underway. I wish he would have mentioned that in his congratulatory letter to AHS and the Project Aries 1 Team.

ABBREVIATIONS

AHS – Anniston High School
ΔV – Delta velocity (velocity change)
ECU – Environmental control unit
EOIB – Earth orbit insertion burn
LH_2 – Liquid Hydrogen
MCCB – Mid-course correction burn
MIT – Massachusetts Institute of Technology
MOIB – Mars orbit insertion burn
MRE – Meals Ready to Eat
NASA – National Aeronautics and Space Administration
Recon – Reconnaissance
SNAP - Systems Nuclear Auxiliary Power Program (SNAP) reactor
TOIB – Transfer orbit insertion burn

REFERENCES

"Anniston Astronauts Are Back Home", Page 1, Anniston Star Newspaper, April 25, 1969.

"How To Do A Science Fair Project", https://www.jpl.nasa.gov/edu/learn/activities/science-fair-project/

Journal of Cosmology; "Interplanetary Trajectory Analysis and Logistical Considerations of Human Mars Exploration"; October-November 2010; MIT/JPL. http://strategic.mit.edu/docs/2_35_JOC_12_3588_Mars_Trajectories.pdf.

Larson, Wiley and Linda Pranke; "Human Spaceflight – Mission Analysis and Design"; McGraw-Hill, New York, NY; ISBN 978-0-07-236811-6.

"Mars Trip At Anniston High Ends", Page 1, Anniston Star Newspaper, April 24, 1969.

"STEAM Portal Page", https://educationcloset.com/steam/what-is-steam/

"Students Make Trip to Mars (Simulated)", Page 1, Anniston Star Newspaper, April 13, 1969.

Sutton, George P.; "Rocket Propulsion Elements", Eighth Edition, John Wiley & Sons, Hoboken, NJ, 2010; ISBN 978-0-470-08024-5.

Tribble, Alan C.; "The Space Environment", Princeton Paperbacks, Princeton University Press, Princeton, NJ; 2003; ISBN 0-691-10299-6.

Von Braun, Dr. Wernher; "The Mars Project" October 1, 1962; University of Illinois Press, Chicago, Il. ISBN 978-0252062278.

Zubrin, Robert; "Entering Space", Penguin Putnam, Inc.; New York, NY. 1999. ISBN 0-87477-975-8.

Zubrin, Robert; "The Case For Mars"; Free Press, New York, NY; 2011; ISBN 978-1-4516-0811-3.

The Anniston Star

AP, UPI, NEA, Los Angeles Times-Washington Post Services   "Your Home Newspaper Since 1882"

64 Pages, 6 Sections   ANNISTON, ALABAMA, SUNDAY, APRIL 13, 1969   10c DAILY, :

Contact Maintained

Communications Engineer and Mission Control Director Jim Young, left, and Head Flight Director Austin maintain constant contact with the astronauts through both radio and video equipment.

The simulated space capsule was built in a storage room at Anniston High School, with Mission Control occupying a downstairs classroom.

The Anniston Star

AP, UPI, NEA, Los Angeles Times-Washington Post Services   "Your Home Newspaper Since 1882"

64 Pages, 6 Sections   ANNISTON, ALABAMA, SUNDAY, APRIL 13, 1969   10c DAILY,

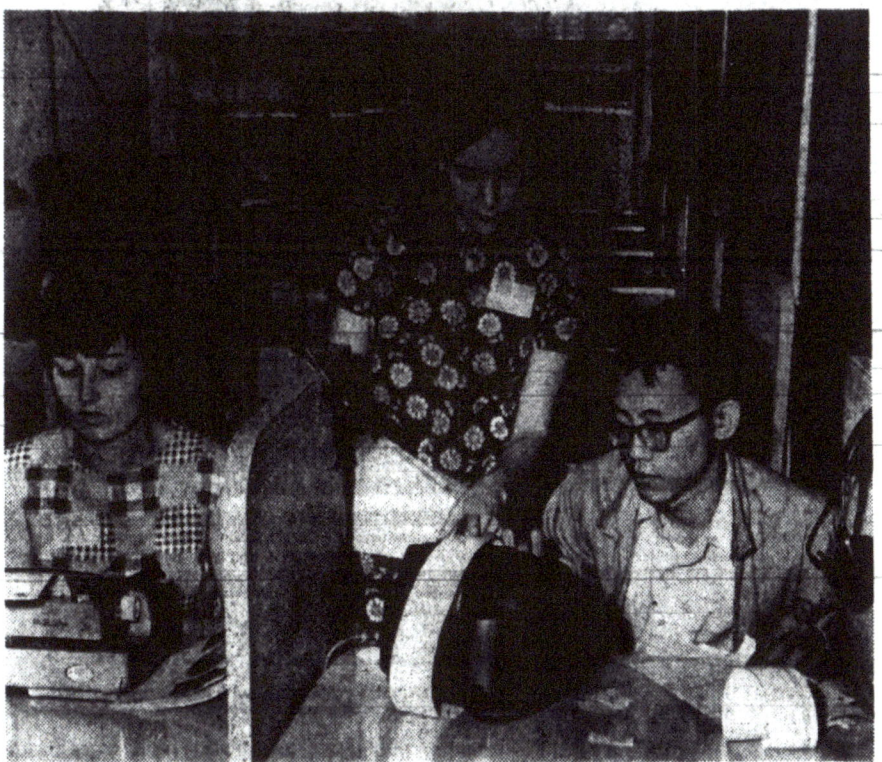

Coordinating Data

Assistant Flight Director Sally Wallace, standing, confers with Ellen Austin, of the medical team, and Ken Mok, head of simulation. The sixteen-day mission, which is now nearing its halfway point, utilizes all the controls necessary in a space capsule — and some devices still in the planning stages at NASA.

Eight Days to Mars - Aries 1

21

Students Make Trip To Mars (Simulated)

.By JUDY JOHNSON

A handful of Anniston High School math and science students have taken on a project too big for even NASA to consider at present.

They're sending two astronauts to Mars.

Well, not really. But the simulation is good enough to make even the most avid follower of the nation's space exploits sit up and take notice.

A storage room at the school has been converted into a two-story simulated space capsule, complete with elaborate electronics and guidance systems, designed and built by the students. And, as principal Clifford Smith says, "It's not just some kid sort of thing where they lock themselves in the closet."

One Grounded

The astronauts are Wayne McCain, 18, a senior, and command pilot, and Wallace Price, 17, a junior. A third astronaut, Sammy Martin, 18, a senior, started the mission with the others, but had to be removed from the capsule by a rescue team when he developed a severe case of intestinal flu.

The sixteen-day mission, which included a blast-off from an orbiting space station, has now been under way more than six days. The spacecraft, Ares I, will orbit Mars several times while the astronauts photograph the surface of the planet in hopes of finding a suitable site for future landings.

The craft is scheduled to reach Mars orbit on the eighth day of the mission. There, it will rendezvous with an unmanned cargo vessel carrying extra supplies. After the rendezvous and the picture-taking session, the craft will leave Mars orbit for the return to the space station.

The reason for the project?

.Stress Studied

As Richard Austin, 17, a senior and flight director of the project, explains it, the flight apparatus is the supporting framework for an experiment in p s y c h o l o g y.

★

Stress is induced 'by t i n y subconscious doubt that maybe it is real.'

★

"We're trying to study how man functions under the stress of space flight." Austin explains that because the astronauts realize the project is only a simulation, the only way they can be placed under a stress similar to that experienced in a real space flight is to "create a tiny subconscious doubt that maybe it is real." For this reason, the equipment in the craft has been made as real as possible.

Understanding such stresses is important, Austin says, because "as NASA's mission becomes more and more that of colonizing other planets, they will have to put more and more stress on getting people there in top physical and mental condition, so they can tackle the problems they'll face in colonizing and living on a hostile planet."

The lower story of the simulated craft c o n t a i n s bunks, kitchen faciliies, lab facilities and a chemical toilet. The upper story houses the control instruments where the simulated functions of the craft are controlled.

.Activities Televised

All of the controls necessary in a space capsule, and some that are still under development by NASA, are being utilized in the simulated vehicle. Complete communication is maintained with mission control, which is located downstairs at the school. C ommunications equipment includes not only radio receiving and transmitting equipment, but television transmitting and receiving equipment. With the aid of a camera inside the capsule, t h e astronauts can give people in mission control a glimpse of their activities during the day.

A mission control team, a security team and a medical team keep a 24-hour per day watch to make c e r t a i n everything goes well with the project.

All of the equipment aboard the simulated craft is functional, according to J i m Young, 17, a junior who serves as systems engineer, mission control director and com-

(See Students, Page 8A, Col. 4)

Eight Days to Mars - Aries 1

Mars Trip At Anniston High Ends

By JUDY JOHNSON

Two Anniston High School students who have spent the past 16 days in a mock-up space capsule on a simulated flight to Mars and back emerged today amid the cheers of their fellow students.

Wayne McCain, 18, and Wallace Price, 17, grinned and waved as they walked from the "capsule" in the high school's chemistry room at 8:30 a.m.

Both had grown heavy beards. They wore orange jump suits and carried their space helmets in their arms.

Greeted With Ovation

The boys were rushed to the school's medical room for a checkup, then to the stage of the school auditorium, where they were greeted with a standing ovation from the student body and faculty assembled there.

Anniston High School Principal Clifford Smith told the astronauts and the audience that all of Anniston "and probably the whole state is aware of this project."

He said the students had received special commendations from Gov. Albert Brewer, the state legislature, Congressman Bill Nichols, NASA and other space agencies.

NASA To Get Data

Richard Austin, flight director of the project, told the assembly that the flight was designed "to study the effects of prolonged space flight on mankind." He said the project group plans to turn the data gathered over to the National Aeronautics and Space Administration.

The astronauts received a second standing ovation from (See Anniston, Page 12A, Col. 7)

ANNISTON

(Continued from Page 1) the student body as they stood to speak. Both thanked the student body for its support and expressed appreciation to the ground crew for its work.

"It didn't seem like 16 days," said Command Pilot Wayne McCain. "It seemed like yesterday we went in, and between then and now is just a big blur," McCain told the assembly that he believed "Project Aries marks the beginning of space research at Anniston High School.

Following the assembly, the boys took their first shower in 16 days, then attended a reception held in their honor at the school cafeteria. The remainder of the day was to be spent in debriefing.

Brazil produces about 4,700,000 tons of steel a year.

Anniston Astronauts Are Back Home

By JUDY JOHNSON

After 16 days of experiencing the induced psychological stresses of an extended space flight, two Anniston High School students have "returned to earth."

The ambitious project, which has gained participating students commendations from Gov. Albert Brewer, the state legislature and NASA, among others, included a simulated flight to Mars and back. The purpose of the flight, according to flight director Richard Austin, 17, was to study how man functions under the stress of space flight.

The actual "flight" began 17 days ago and ended Thursday morning. But the project itself got underway several months ago. A storage room at the school was converted into a two-story simulated space capsule, complete with elaborate electronics and guidance systems, designed and built by students in the project.

Lived In Capsule

Two boys — Wayne McCain, 18, and Wallace Price, 17, were chosen as the project's astronauts, and actually lived in the capsule more than two weeks. But over 40 students participated in the project in various capacities.

A third astronaut, Sammy Martin, 18, started the mission with the other two, but had to be removed from the capsule by a rescue team when he developed intestinal flu.

McCain said Thursday that after Martin became sick, the other two underwent emergency procedures to prevent infection. "We were hoping we wouldn't catch the virus," McCain said. They didn't.

"We never doubted we would make it," McCain said. "We kept our minds occupied with the technical parts of the flight."

Long Flights Possible

What was learned during the project?

"Wally and I learned that it is totally possible at this time to have long-term space flights," McCain said. "Our technology and advancements made in human relations make it possible." But, McCain added, "without the people on the ground, we couldn't have done it. I know they had a rougher time than we did."

"From what we have learned," flight director Austin said, "if an astronaut has something to do, he can take it pretty well. If he has little to do, stress starts showing in his actions, and he becomes short - tempered. We kept them busy, and they kept themselves busy writing down logs and doing other tasks."

Austin said that apart from the illness early in the flight and "a few malfunctions in systems" which were minor and quickly repaired, no trouble was encountered during the mission.

Greeted With Cheers

The astronauts were greeted by cheering students as they stepped from their capsule at 8:30 a.m. Thursday. Both had grown heavy beards. They wore orange jump suits and cradled their space helmets in their arms. They were immediately taken to the school's medical room where they were checked over, and then to an assembly of the school's student body in the auditorium, where they received a standing ovation and certificates of merit from the student body.

Following the assembly the astronauts showered, then attended a reception in their honor in the school cafeteria.

Debriefing was scheduled to occupy the rest of the day, "and probably on into the night," according to Austin.

During the debriefing the astronauts were to give their ideas and impressions about how they felt during the flight, as well as data on the performance of the systems aboard the space capsule. The data will be analyzed and a final report will be prepared which will include any

(See Home, Page 6A, Col. 1)

The Anniston Star

AP, UPI, NEA, Los Angeles Times-Washington Post Services

"Your Home Newspaper Since 1882"

VOL. 87, NO. 219   ANNISTON, ALABAMA, FRIDAY, APRIL 25, 1969   14 Pages, 2 Sections

(McElroy Photo)

Mission Complete

Anniston High School astronauts Wallace Price, left, and Wayne McCain relax and admire each other's beards after 16 days in a mock-up space capsule on a flight to Mars. The Aries I astronauts were later greeted by a standing ovation from the entire school student body.

Eight Days to Mars - Aries 1

24

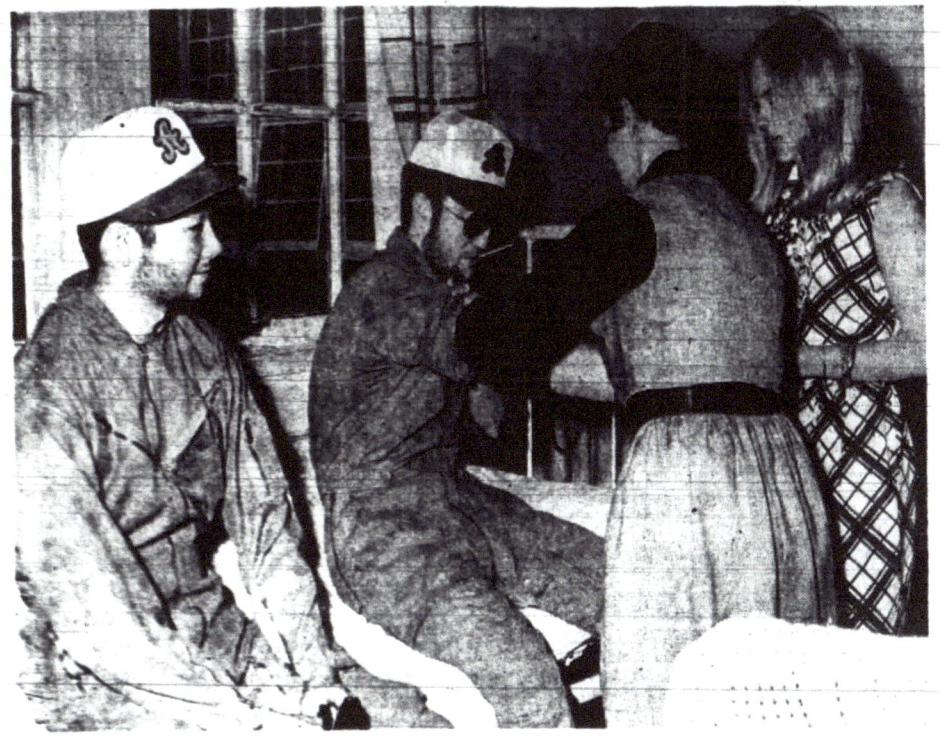

CONDITION CHECKED — Aries I Command Pilot Wayne McCain has his blood pressure checked while Wallace Price, left, waits his turn after the two completed their 16-day simulated flight to Mars Thursday. Following an assembly and a short reception, the two were scheduled to spend the remainder of the day in debriefing. Data from the experiment will be used in a report, one copy of which will be sent to NASA. (McElroy Photo)

ABOUT THE AUTHOR

Dr. J. Wayne McCain has been a professional project manager in the DoD/aerospace industry for over 30 years of that time. Dr. McCain graduated from Auburn University with a Bachelor of Science in Aerospace Engineering and has MBA and PhD (in engineering) degrees from the University of Alabama in Huntsville. In May 2016, he earned a Master of Space Systems engineering from Florida Institute of Technology where he headed a group thesis project being considered for publication.

It concentrated on risk analyses of humans-to-Mars space missions along with mission planning. McCain has worked Army and Air Force related defense programs for Thiokol Propulsion and Martin Marietta (now Lockheed Martin) in Denver. Dr. McCain also worked on automated test equipment for the NASA Space Shuttle Program while at United Space Boosters, Inc. He has served as Principal Investigator for several defense-related Small Business Innovative Research (SBIR) endeavors and has a recent patent for a laser initiated ordnance system (LIOS) safe-arm device. He has received recognition from the USAMRDEC at Redstone for assistance in missile development and fault analysis while acting as a Engineering and Technical Consultant to the Army. Dr. McCain has been involved in college education since the late 1970's and has witnessed first-hand the influences (both good and bad) had by evolution of the 'information age', personal computers, and the Internet. For over 10 years, he served as Chair of the Management of Technology (MOT) Department within the College of Business and developed the Technology Transfer and Risk Management courses that are taught as part of the MOT Degree. He developed the Aviation Management, MOT, and Project Management Certificates and Minors, and a 'Tailored Minor' for COB students. Dr. McCain is an Assoociate Fellow of the American Institute of Aeronautics and Astronautics (AIAA), a member of the IEEE (Institute of Electrical and Electronic Engineering), PMI (Project Management Institute), and SARA (Society of Amateur Radio Astronomers) where he is a Director and a participant in the JOVE and SuperSID Programs, NASA-sponsored Radio Astronomy STEM (Science, Technology, Engineering, and Math) projects. He has proposed to develop and launch a Radio Astronomy 'CubeSat' to support STEM in K-12 and colleges and is working with the STEM-SAT organization to submit a proposal to NASA for a launch slot. He is also a former Chair of the Alabama Academy of Science (AAS) STEM Section Committee and presided over the STEM paper and poster competitions at the February 20-22, 2019 AAS Annual meeting held at the historical Tuskegee University. He is faculty advisor for the Athens State AIAA Student Branch and the College of Business Internet Radio Station 'KASUradio.net' (also heard on 89.9 FM campus-wide). He most recently completed training with the National Weather Service as a SKYWARN storm spotter in conjunction with the Emergency Management support provided by the Athens State Amateur Radio Club Station, W4CQD where McCain serves as the FCC-designated trustee and chief operator. He has established a MOT Laboratory that uses 3-D printing as one of the example 'high-tech' management challenges for students. McCain will also help coordinate the 99th Annual Meeting of the Alabama Academy of Science to be hosted by Athens State in March of 2022.

Second Printing 7-2019
J. Wayne McCain

ISBN: 9781081668877

Eight Days to Mars - Aries 1

www.ingramcontent.com/pod-product-compliance
Lightning Source LLC
Chambersburg PA
CBHW041120180526
45172CB00001B/340